M336
Mathematics and Computing: a third-level course

GROUPS & GEOMETRY

UNIT GR6
FINITE GROUPS 2

Prepared for the course team by
Bob Coates & Bob Margolis

The Open University

This text forms part of an Open University third-level course.
The main printed materials for this course are as follows.

Block 1
Unit IB1   Tilings
Unit IB2   Groups: properties and examples
Unit IB3   Frieze patterns
Unit IB4   Groups: axioms and their consequences

Block 2
Unit GR1   Properties of the integers
Unit GR2   Abelian and cyclic groups
Unit GE1   Counting with groups
Unit GE2   Periodic and transitive tilings

Block 3
Unit GR3   Decomposition of Abelian groups
Unit GR4   Finite groups 1
Unit GE3   Two-dimensional lattices
Unit GE4   Wallpaper patterns

Block 4
Unit GR5   Sylow's theorems
Unit GR6   Finite groups 2
Unit GE5   Groups and solids in three dimensions
Unit GE6   Three-dimensional lattices and polyhedra

The course was produced by the following team:

Andrew Adamyk (BBC Producer)
David Asche (Author, Software and Video)
Jenny Chalmers (Publishing Editor)
Bob Coates (Author)
Sarah Crompton (Graphic Designer)
David Crowe (Author and Video)
Margaret Crowe (Course Manager)
Alison George (Graphic Artist)
Derek Goldrei (Groups Exercises and Assessment)
Fred Holroyd (Chair, Author, Video and Academic Editor)
Jack Koumi (BBC Producer)
Tim Lister (Geometry Exercises and Assessment)
Roger Lowry (Publishing Editor)
Bob Margolis (Author)
Roy Nelson (Author and Video)
Joe Rooney (Author and Video)
Peter Strain-Clark (Author and Video)
Pip Surgey (BBC Producer)

With valuable assistance from:

Maths Faculty Course Materials Production Unit
Christine Bestavachvili (Video Presenter)
Ian Brodie (Reader)
Andrew Brown (Reader)
Judith Daniels (Video Presenter)
Kathleen Gilmartin (Video Presenter)
Liz Scott (Reader)
Heidi Wilson (Reader)
Robin Wilson (Reader)

The external assessor was:
Norman Biggs (Professor of Mathematics, LSE)

The Open University, Walton Hall, Milton Keynes, MK7 6AA.

First published 1994. Reprinted 1997, 2002, 2005, 2007.

Copyright © 1994 The Open University

All rights reserved. No part of this publication may be reproduced, stored in a retrieval system or transmitted in any form or by any means, without written permission from the publisher or a licence from the Copyright Licensing Agency Limited. Details of such licences (for reprographic reproduction) may be obtained from the Copyright Licensing Agency Ltd of 90 Tottenham Court Road, London, W1P 9HE.

Edited, designed and typeset by the Open University using the Open University TEX System.

Printed in Malta by Gutenberg Press Limited.

ISBN 0 7492 2168 2

This text forms part of an Open University Third Level Course. If you would like a copy of *Studying with the Open University*, please write to the Central Enquiry Service, PO Box 200, The Open University, Walton Hall, Milton Keynes, MK7 6YZ. If you have not already enrolled on the Course and would like to buy this or other Open University material, please write to Open University Educational Enterprises Ltd, 12 Cofferidge Close, Stony Stratford, Milton Keynes MK11 1BY, United Kingdom.

CONTENTS

Study guide	4
Introduction	5
1 Review	5
2 Groups of order $2p$	11
3 Groups of order 12 (audio-tape section)	15
4 Practice exercises	27
5 Where now?	28
5.1 Composition series	28
5.2 Soluble and simple groups	30
5.3 Nilpotent groups	31
5.4 Representations	31
5.5 Conclusion	32
Solutions to the exercises	33
Objectives	42
Index	43

STUDY GUIDE

This unit contains no essentially new material. The first section should take approximately one-fifth of the study time for the unit. The second section is rather shorter, while the third is rather longer. The remaining two sections are of similar length to the first.

The fourth section is of particular importance as it contains examples of questions designed to test your understanding of the basic ideas from the Groups stream. It is specifically designed to help you in your revision for the examination.

There is an audio programme associated with Section 3.

INTRODUCTION

The aims of this unit are to review and apply the major results that we have obtained in the Groups stream.

In Section 1 we gather together these results, with some typical applications. We also revisit the original definition of group action and link it to permutation groups.

Section 2 applies the ideas from Section 1 to the class of groups of order $2p$, where p is an odd prime. We obtain a classification of all such groups.

Section 3, which is quite long, classifies all groups of order 12. Besides the actual results, this section illustrates how the general theorems can be applied, giving useful revision.

Further revision and practice is provided in Section 4.

The final section of the unit indicates how some of the main ideas discussed in the Groups stream can be generalized.

1 REVIEW

In this section we shall look at the methods that we have used so far in the Groups stream and at the classification results that these methods have enabled us to prove. We shall also revisit the original definition of group action in order to provide a link with permutation groups. It is this link with permutation groups that will be useful in our classification of groups of order 12 in Section 3. In our review, we give brief marginal references for major ideas to the appropriate earlier units.

The concepts of free groups and free Abelian groups have underpinned our whole approach to defining groups in terms of generators and relations. The existence of these free groups guarantees that any finite presentation that we care to write down will define a group. Despite knowing that a presentation must define a group, we may well not be able to decide the order of the group or, indeed, whether the group is finite.

The group defined will be a quotient of a free group by the normal subgroup generated by the relations of the presentation.

In the case of Abelian groups, the situation is much more satisfactory. In fact, we proved a classification theorem for all finitely generated Abelian groups.

Unit GR3

Theorem *Canonical decomposition of finitely generated Abelian groups*

If A is a non-trivial Abelian group generated by a finite number of elements then

$$A \cong \mathbb{Z}_{d_1} \times \cdots \times \mathbb{Z}_{d_k},$$

where

$$d_i \in \mathbb{Z}, \quad d_i > 1 \text{ or } d_i = 0, \quad d_i \mid d_{i+1}, i = 1, \ldots, k-1.$$

The trivial Abelian group, $\{0\}$, is the trivial cyclic group $\mathbb{Z}_1$.

So our knowledge of finitely generated Abelian groups depends on our knowledge of the group of integers $\mathbb{Z}$ under addition and of its quotient groups $\mathbb{Z}_n$.

For a finitely *presented* Abelian group, that is where we also have a finite
number of relations, we have an algorithm for determining the group in
canonical form. This algorithm, referred to as the Reduction Algorithm, *Unit GR3*
involves reducing the matrix, corresponding to the generators and relations,
to diagonal form via a series of elementary row and column operations.

For groups which we do not know to be Abelian, we have proved rather
more restricted results. These have either been about all groups of a specific
order, for example all groups of order 8, or about a whole class of groups *Unit GR2*
whose orders have some more general restriction. For example, we know
that groups of prime order must be cyclic and groups whose order is the
square of a prime p must be either the Abelian group $\mathbb{Z}_p \times \mathbb{Z}_p$ or the *Unit GR4*
Abelian group $\mathbb{Z}_{p^2}$.

We shall consider two further classifications in this unit. In Section 2 we
consider the class of groups whose order is twice a prime and in Section 3 we
classify all groups of order 12.

Our main tools in analysing groups in general have been the Sylow
theorems. These theorems tell us of the existence of subgroups of maximal *Unit GR5*
prime power orders. They also place restrictions on the numbers of such
subgroups. When these restrictions tell us that there is a unique, and hence In *Unit GR5* we only discussed the
normal, Sylow subgroup corresponding to each prime, then we can express case where the order of the group
the group as an internal direct product of its Sylow subgroups. is $p^m q^n$, i.e. where there is a
This may still not determine the group uniquely. For example, if all we know unique Sylow subgroup for each of
is that the Sylow 2-subgroups of a group have order 8, then we know that *two* primes. In general, if we have
there are five possibilities for the structure of these subgroups. Knowing the a unique Sylow subgroup for each
order of a Sylow p-subgroup does not necessarily determine its structure. prime dividing the order of a
group, then these subgroups will be
normal and we can use an extended
The general study of groups grew, historically, from the study of version of the Internal Direct
permutation groups. The link between finite groups and permutation groups Product Theorem to express the
is provided by Cayley's Theorem (*Unit GR4*). This states that any group of group as an internal direct product
order n is isomorphic to a subgroup of S_n. So, for example, to classify of its Sylow subgroups.
groups of order 12 all we need to do is investigate all the subgroups of
order 12 in S_{12}. We noted, however, that Cayley's Theorem was of
theoretical rather than practical use, since S_{12} is a group of order

$$12! = 479001600,$$

and so to investigate all its subgroups of order 12 would be a major
undertaking.

In practice, what is much more useful than Cayley's Theorem is to establish
that a group G is isomorphic to a subgroup of S_n where n is much smaller
than the order of G. The Sylow theorems and group actions often enable
this to be done.

Our proofs of the Sylow theorems depend heavily on the use of group
actions and, in fact, as we shall demonstrate below, *any* group action
provides a permutation group homomorphic to the original group.

The definition of an action of a group G on a set X that we have been using
throughout the Groups stream is expressed in terms of the operation $\wedge$
which satisfies the following conditions:

(a) $g \wedge x \in X$, for all $g \in G, x \in X$;

(b) $e \wedge x = x$, for all $x \in X$, where e is the identity element of G;

(c) $(gh) \wedge x = g \wedge (h \wedge x)$, for all g, h in G and $x \in X$.

On the other hand, our original definition of group action involved groups of
permutations, and it is this definition which enables us to associate a
subgroup of some permutation group S_n with every group action.

The original definition of group action was as follows.

Definition Group action

A **group action** consists of a group G, a set X and a mapping
$$\phi : G \to \Gamma(X)$$
$$g \mapsto \phi_g$$
from G to the group $\Gamma(X)$ of bijections from X to X, such that
$$\phi_{gh} = \phi_g \circ \phi_h, \quad \text{for all } g, h \in G.$$

Bijections are functions which are one–one and onto.

The relation between ϕ and the corresponding $\wedge$ is given by
$$g \wedge x = \phi_g(x), \quad g \in G, x \in X.$$

In normal functional notation, we would have used $\phi(g)$ rather than the historical ϕ_g for the bijection associated with g. When we do so, the condition
$$\phi_{gh} = \phi_g \circ \phi_h$$
becomes
$$\phi(gh) = \phi(g) \circ \phi(h), \quad \text{for all } g, h \in G.$$

Thus, this definition says that ϕ is a homomorphism from the group G to the group $\Gamma(X)$ of bijections of X.

Now the permutation group S_n is defined to be the set of bijections of the particular set (of n elements)
$$\{1, \ldots, n\}.$$

However, if the set X in a group action has n elements, say
$$X = \{x_1, \ldots, x_n\},$$
then the set of all bijections of X (with composition of functions as its operation) is a group isomorphic to S_n. Informally, if a permutation of S_n maps i to j, then the corresponding bijection of X maps x_i to x_j. Conversely, given a bijection of X, we can write down the corresponding permutation of the suffices to get an element of S_n.

Now suppose that the group G acts on the set
$$X = \{x_1, \ldots, x_n\}.$$

The function ϕ maps each element g of G to a bijection of X, which is then mapped by the above isomorphism to a corresponding permutation in S_n.

Thus, formally, we have a composite map
$$G \to \Gamma(X) \to S_n.$$

Since the first function ϕ is a homomorphism and the second is an isomorphism, the composite from G to S_n is a homomorphism. This is the homomorphism that we shall refer to as being *derived* from the group action.

Rather less formally, we shall identify the elements $x_1, \ldots, x_n$ with $1, \ldots, n$ and say that a group action of G on a set with n elements gives a homomorphism ϕ from G to S_n.

Now the First Isomorphism Theorem tells us that
$$G/\operatorname{Ker}(\phi) \cong \operatorname{Im}(\phi).$$

Unit IB4

This isomorphism is most useful when the kernel is trivial, i.e. when ϕ is one–one, since in that case the group G is *isomorphic* to $G/\operatorname{Ker}(\phi)$ and hence to the subgroup $\operatorname{Im}(\phi)$ of S_n.

Unfortunately, the homomorphism ϕ may or may not be one–one, so it is not always the case that a group action provides a subgroup of S_n *isomorphic* to G. To decide whether or not we have an isomorphism requires us to decide whether or not the kernel is trivial.

In the case where we do have an isomorphism, and if n is less than the order of G, we have that G is isomorphic to a subgroup of a permutation group S_n much smaller than that given by Cayley's Theorem.

To illustrate the discussion above, we now look at the particular example of a group G of order 22. If G is Abelian then, since $22 = 2 \times 11$, we know that

$$G \cong \mathbb{Z}_2 \times \mathbb{Z}_{11} \cong \mathbb{Z}_{22}.$$

Unit GR2

In the following exercise we ask you to begin to investigate the non-Abelian case.

Exercise 1.1

Let G be a non-Abelian group of order 22.
Show that G has one Sylow 11-subgroup and eleven Sylow 2-subgroups.

Now let

$$X = \{H_1, \ldots, H_{11}\}$$

be the set of Sylow 2-subgroups of G. Since the conjugate of any two-element subgroup of G is again a two-element subgroup of G, we know that

$$g \wedge H_i = gH_ig^{-1}, \quad g \in G, H_i \in X,$$

defines an action of G by conjugation on X. Furthermore, by the Sylow theorems, all the Sylow 2-subgroups are conjugate, and as a consequence this action has just one orbit.

By the previous discussion, the action of G on X defined by $\wedge$ gives rise to a corresponding homomorphism ϕ from G to S_{11}.

We now consider the kernel of ϕ. For an element g in G to belong to the kernel, it must act as the identity permutation on X, that is it must fix every element of X. In the language of group actions this says that g must belong to the stabilizer of every element of X. This remark is perfectly general and worth recording as a result in its own right.

Theorem 1.1

Let G be a group acting on a set X of size n and let ϕ be the homomorphism from G to S_n derived from the group action. Then

$$\mathrm{Ker}(\phi) = \bigcap_{x \in X} \mathrm{Stab}(x).$$

So, for our group of order 22, in order to find $\mathrm{Ker}(\phi)$, we first find the stabilizer of a typical element H_i of X. Now, as we have already observed, the Sylow theorems tell us that the orbit of any H_i consists of all the H_is. So

$$|\mathrm{Orb}(H_i)| = 11.$$

The Orbit–stabilizer Theorem tells us that

$$|G| = |\mathrm{Orb}(H_i)| \times |\mathrm{Stab}(H_i)|$$

and, since $|G| = 22$ and $|\mathrm{Orb}(H_i)| = 11$, we have

$$|\mathrm{Stab}(H_i)| = 2.$$

We now make use of the fact that the action of G on X was defined by conjugation. That is, if g is in G, then

$$g \wedge H_i = gH_ig^{-1}.$$

If h is an element of H_i, it follows, by the closure property of subgroups, that

$$h \wedge H_i = hH_ih^{-1} = H_i,$$

giving $h \in \text{Stab}(H_i)$. So

$$H_i \subseteq \text{Stab}(H_i).$$

However, we know that both H_i and $\text{Stab}(H_i)$ have two elements, so

$$\text{Stab}(H_i) = H_i.$$

The kernel of ϕ, being the intersection of the stabilizers, is therefore the intersection of all the H_is. However, we can use Lagrange's Theorem to deduce that two distinct subgroups each of prime order have trivial intersection, so

Theorem 1.1

$$\text{Ker}(\phi) = \bigcap_{H_i \in X} \text{Stab}(H_i) = \bigcap_{H_i \in X} H_i = \{e\}.$$

Since $\text{Ker}(\phi) = \{e\}$, what we have shown is that G is *isomorphic* to a subgroup of S_{11}. Although S_{11} is still a large group, it is much smaller than S_{22} as provided by Cayley's Theorem.

Before showing that we can completely determine the structure of any group G of order 22, let us review what we have done so far.

(a) We used the Sylow theorems to determine the number of Sylow subgroups corresponding to each prime. This information was particularly useful in this case because there was only one possibility for a non-Abelian group.

(b) We looked at the action of the group by conjugation on the set of Sylow subgroups belonging to a particular prime and at its corresponding homomorphism.

(c) Because the action is conjugation, we saw that the Sylow theorems tell us that there is a single orbit. This made the stabilizers easier to determine.

(d) We saw that the kernel of the homomorphism derived from the group action is the intersection of the stabilizers. The task of finding this intersection was made easier by the fact that the Sylow subgroups had prime order.

To complete the analysis of the structure of any group G of order 22, we look at the unique and hence *normal* Sylow 11-subgroup K.

Let K have generator a and pick any element b of order 2, i.e. a generator of any one of the Sylow 2-subgroups. Because K is normal in G, we know that

$$bab^{-1} \in K.$$

However,

$$K = \langle a \rangle = \{e = a^0, a^1, \ldots, a^{10}\},$$

and so

$$bab^{-1} = a^i$$

for some $i = 0, 1, \ldots, 10$. We can immediately rule out the case where $i = 0$ since this would lead to the conclusion that $a = e$, contradicting the fact that a has order 11. Therefore, $i = 1, \ldots, 10$.

Conjugating again by b, we have

$$b(bab^{-1})b^{-1} = ba^ib^{-1} = (bab^{-1})^i = (a^i)^i = a^{i^2}.$$

Since $b^2 = e$, we have
$$b(bab^{-1})b^{-1} = a = a^{i^2},$$
and so
$$a^{i^2-1} = e.$$

Now a has order 11 and so it follows that
$$11 \mid (i^2 - 1), \quad \text{that is} \quad 11 \mid (i-1)(i+1).$$

Since 11 is prime this implies that either 11 divides $i - 1$ or that 11 divides $i + 1$. Combining this with the fact that $i = 1, \ldots, 10$ shows that the only possibilities are
$$i = 1 \quad \text{and} \quad i = 10.$$

Now the subgroup K has index 2 (it has 11 elements) and the two left cosets $eK = K$ and bK are distinct (bK contains the element b of order 2 whereas the elements of K are of order 1 or 11). This information shows that a and b generate G, because the cosets K and bK, all of whose elements are powers of a and b, account for all 22 elements of G.

We also know that
$$bab^{-1} = a \quad \text{or} \quad bab^{-1} = a^{10} = a^{-1}.$$

The first possibility says that the generators of G commute, leading to the Abelian group
$$\mathbb{Z}_2 \times \mathbb{Z}_{11} \cong \mathbb{Z}_{22}$$
considered earlier. Thus, if a non-Abelian group of order 22 exists then it has an element a of order 11 and an element b of order 2 such that $bab^{-1} = a^{10}$, i.e. such that $ba = a^{10}b = a^{-1}b$. In other words, it has a presentation
$$\langle a, b \colon a^{11} = e,\ b^2 = e,\ ba = a^{10}b\ (= a^{-1}b)\rangle.$$

There is such a group: the dihedral group D_{11}. Thus we have proved that any non-Abelian group of order 22 must be the dihedral group D_{11}.

We can conclude therefore that any group of order 22 is either the cyclic group $\mathbb{Z}_{22}$ or the dihedral group D_{11}.

We shall generalize this argument in Section 2, where we classify groups of order $2p$, where p is any odd prime.

This classification of groups of order 22 did not use the isomorphism which we found existed from the group to a subgroup of S_{11}, since other techniques enabled us to describe the group completely. In general, however, the homomorphism associated with conjugation of Sylow subgroups may, by itself, enable us to classify the group. Indeed, in this case, the existence of this isomorphism enables us to give a second concrete example of the non-Abelian group of order 22 by looking inside S_{11}.

Because G must have an element of order 11, so must $\text{Im}(\phi)$. The only elements of order 11 in S_{11} are the 11-cycles. We can try the 11-cycle
$$c = (1\,2\,3\,4\,5\,6\,7\,8\,9\,10\,11)$$

for the element corresponding to the generator a. Next we want an element of S_{11} which conjugates the above 11-cycle to give its inverse
$$c^{-1} = (1\,11\,10\,9\,8\,7\,6\,5\,4\,3\,2).$$

As we have seen previously, there is an easy way to find an element conjugating one permutation to another of the same cycle type. We write the second permutation below the first with the individual cycle components in corresponding positions. A permutation conjugating the first permutation to the second is one which maps the elements in the first permutation to the corresponding elements in the second. Hence, with the two 11-cycles written as above we obtain

$$d = \begin{pmatrix} 1 & 2 & 3 & 4 & 5 & 6 & 7 & 8 & 9 & 10 & 11 \\ 1 & 11 & 10 & 9 & 8 & 7 & 6 & 5 & 4 & 3 & 2 \end{pmatrix}.$$

Written in cycle form this is

$$d = (1)(2\,11)(3\,10)(4\,9)(5\,8)(6\,7).$$

Since d is the product of disjoint 2-cycles we have

$$d^2 = e$$

and, by the way it was chosen,

$$dcd^{-1} = c^{-1}.$$

The permutations c and d generate a non-Abelian subgroup of S_{11} of order 22.

The connection between this subgroup of S_{11} and D_{11} is easiest to see if you number the vertices of a regular 11-gon from 1 to 11 anticlockwise, as in Figure 1.1. The 11-cycle then corresponds to an anticlockwise rotation through $2\pi/11$ and the permutation d to reflection in the axis of symmetry through vertex 1.

The following exercise gives you an opportunity to apply arguments similar to those employed above.

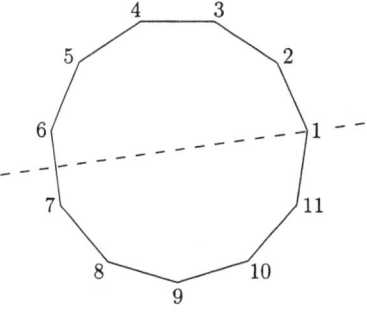

Figure 1.1

Exercise 1.2

Let G be a group of order 405.

(a) Show that G has a unique Sylow 3-subgroup and either 1 or 81 Sylow 5-subgroups.

Consider the case of 81 Sylow 5-subgroups. Let

$$X = \{H_1, \ldots, H_{81}\}$$

be the set of Sylow 5-subgroups of G and let G act on X by conjugation.

(b) Show that

$$\text{Stab}(H_i) = H_i, \quad i = 1, \ldots, 81.$$

(c) Deduce that the homomorphism

$$\phi : G \to S_{81}$$

derived from the action is one–one and hence that G is isomorphic to a subgroup of S_{81}.

2 GROUPS OF ORDER $2p$

In this section we shall classify all groups of order $2p$ where p is an odd prime. We shall also consider, to some extent, how the techniques used might be applied to the classification of groups of order qp where q and p are distinct primes with $q < p$. We already have examined several special cases. For example, there are two groups of order $6 = 2 \times 3$, namely $\mathbb{Z}_6$ and $S_3 \cong D_3$, and just one group of order $15 = 3 \times 5$, namely $\mathbb{Z}_{15}$. Also, from Section 1, there are only two groups of order $22 = 2 \times 11$, namely $\mathbb{Z}_{22}$ and D_{11}.

We have already dealt with the case where $p = 2$: the only possible groups are $V \cong \mathbb{Z}_2 \times \mathbb{Z}_2$ and $\mathbb{Z}_4$.

The only reason that we specify that q and p be distinct is that we have already classified groups of order p^2. They are either $\mathbb{Z}_{p^2}$ or $\mathbb{Z}_p \times \mathbb{Z}_p$.

The general order qp problem is slightly too complicated to solve completely in this course. However, the order $2p$ case is relatively simple, being an immediate generalization of the order 22 case.

Exercise 2.1

Let G be a group of order $2p$, where p is an odd prime.

(a) Classify the Sylow 2-subgroups and Sylow p-subgroups.

(b) How many Sylow p-subgroups does G possess?

Exercise 2.2

Generalize the result of the previous exercise to a group G of order qp where q and p are primes with $q < p$.

We now know that, for groups of order $2p$, where p is an odd prime, there is a unique, and hence normal, subgroup of order p (isomorphic to $\mathbb{Z}_p$). We shall denote this by K and take one of its generators to be a. In other words,

$$K = \langle a \rangle, \quad \text{where } |a| = p \text{ and } K \text{ is normal in } G.$$

In addition, we know that there is at least one Sylow 2-subgroup of order 2 (isomorphic to $\mathbb{Z}_2$). Let H be one such subgroup, with generator b. In other words,

$$H = \langle b \rangle, \quad \text{where } |b| = 2.$$

Note that the only change we need to make when the group G has order qp with $q < p$ is that, for the Sylow q-subgroup H, we have

$$H = \langle b \rangle, \quad \text{where } |b| = q.$$

We next consider what we can say regarding the number of Sylow 2-subgroups in a group G of order $2p$.

Exercise 2.3

(a) Let G be a group of order $2p$. What are the possible numbers of Sylow 2-subgroups?

(b) By considering the groups of order 6, show that both possibilities can arise.

Before we move on, it is worth recording as a general result the fact that Abelian groups have unique Sylow subgroups, a fact that we can deduce from the alternative part of our solution to Exercise 2.3(b).

> **Lemma 2.1**
>
> Let A be an Abelian group and p be a prime divisor of the order of A. Then A has a unique (normal) Sylow p-subgroup.

In fact the situation is virtually identical for groups of order qp.
We know that the number of Sylow q-subgroups is a divisor of the prime p. As a consequence there is either one such subgroup or p of them.
If there is only one such subgroup then the group has a unique Sylow q-subgroup and, by Exercise 2.2, a unique Sylow p-subgroup. So these are *normal* subgroups of orders q and p. Therefore, in the case where there is only one subgroup of order q, we know that the group must be the internal direct product of its Sylow subgroups. In other words it is $\mathbb{Z}_q \times \mathbb{Z}_p \cong \mathbb{Z}_{qp}$. Theorem 4.1, *Unit GR5*.

Exercise 2.4

Classify all groups of the following orders:

(a) 15;

(b) 77.

Formalizing the comment preceeding Exercise 2.4 gives the following result.

> **Theorem 2.1**
>
> Let G be a group of order qp, where q and p are distinct primes with $q < p$.
> Then, if p is not congruent to 1 modulo q,
>
> $$G \cong \mathbb{Z}_q \times \mathbb{Z}_p \cong \mathbb{Z}_{qp}.$$

We now continue with our investigation of groups of order $2p$. As above, we denote by K the unique normal Sylow p-subgroup. Letting a be some fixed generator of K, we have

$$K = \langle a \rangle, \quad \text{where } |a| = p.$$

In what follows, you might like to note the use of techniques applied to the order 22 case in Section 1.

We choose one of the Sylow 2-subgroups to be H with generator b, in other words,

$$H = \langle b \rangle, \quad \text{where } |b| = 2.$$

It is at this stage that we use the fact that K is a normal subgroup of G. Since this is the case we know that

$$bKb^{-1} \subseteq K$$

and, in particular, that

$$bab^{-1} \in K.$$

However,

$$K = \{e = a^0, a, a^2, \ldots, a^{p-1}\},$$

which leads to the conclusion that

$$bab^{-1} = a^i, \quad \text{for some } i = 0, 1, \ldots, p-1.$$

Exercise 2.5

Explain why the case $i = 0$ can be eliminated.

Exercise 2.6

Use the facts that

$$bab^{-1} = a^i, \quad \text{for some } i = 1, \ldots, p-1,$$

and that $|b| = 2$ to prove that

$$a = a^{i^2}.$$

Exercise 2.7

From the previous exercise, and the fact that $|a| = p$, show that

$$i = 1 \quad \text{or} \quad i = p - 1.$$

Exercise 2.8

Deduce that the only groups of order $2p$, where p is an odd prime, are $\mathbb{Z}_{2p}$ and D_p.

To finish this section we look briefly at the problem of classifying groups of order $3p$ where p is a prime with $3 < p$.

The cases of order $3 \times 2 = 6$ and 3×3 have already been dealt with.

Following our previous work, we observe that the number of Sylow p-subgroups of such a group G divides 3 and is congruent to 1 modulo p. It follows that there is a unique normal Sylow p-subgroup isomorphic to $\mathbb{Z}_p$. Let this subgroup be K where

$$K = \langle a \rangle, \quad \text{where } |a| = p.$$

The number of Sylow 3-subgroups divides p and is congruent to 1 modulo 3. Now if p is not congruent to 1 modulo 3 there is a unique Sylow 3-subgroup and, by the Internal Direct Product Theorem,

$$G \cong \mathbb{Z}_3 \times \mathbb{Z}_p \cong \mathbb{Z}_{3p}.$$

Since any number is congruent to 0, 1 or 2 modulo 3, the primes greater than 3 are congruent to either 1 or 2 modulo 3. The first few primes greater than 3 are

$$5, 7, 11, 13, 17, 19, 23, 29, 31, \ldots$$

and of these 5, 11, 17, 23 and 29 are congruent to 2 modulo 3. Hence there is only one group of each of the orders 15, 33, 51, 69 and 87: the cyclic ones.

This gives rise to the interesting number-theoretic question as to whether there are infinitely many primes congruent to 2 modulo 3, for which orders there is only one group. More generally one can ask the question whether, given two coprime positive integers m and n, there are infinitely many primes congruent to m modulo n.

These number-theoretic problems are discussed in An Introduction to the Theory of Numbers by G.H. Hardy and E.M. Wright (Oxford University Press, 4th edition 1962) and in Elementary Number Theory by D.M. Burton (Allyn and Bacon, revised printing 1980), under the title Dirichlet's Theorem.

There also remains the question of what we can say about groups of order $3p$ when p is a prime greater than 3 which is congruent to 1 modulo 3.

More generally, what is known of groups of order qp where q and p are primes with $q < p$?

In fact the answers to all of these questions are known, and we hope that you find the questions sufficiently intriguing for you to continue your study of group theory beyond this course.

The bibliography given in the Course Guide indicates some possible further reading about groups.

3 GROUPS OF ORDER 12 (AUDIO-TAPE SECTION)

In the previous section we classified all groups whose orders had a particular form. In this section we tackle a more restricted classification problem, namely that of all groups of the particular order 12. This turns out to be more involved than the classification of groups of order $2p$ and gives some indication of the extreme difficulty of the general problem of classifying groups of order n, for arbitrary n. In fact, for some specific values of n, the classification problem is currently unsolved, whereas some whole classes of orders are very simple to deal with.

Those of order twice a prime.

We have classified all groups of order p, p^2 and $2p$, for p prime.

We now review what we know about groups of order 12.

The Abelian groups of order 12 can be found by using the Canonical Decomposition Theorem. Since

$$12 = 2^2 \times 3,$$

the only possibilities are

$\mathbb{Z}_{12}$ and $\mathbb{Z}_2 \times \mathbb{Z}_6$.

We discussed two other groups of order 12 earlier in the course: the dihedral group D_6 and the alternating group A_4.

The dihedral group D_6 can be thought of either as

$$\langle a, b : a^6 = e,\ b^2 = e,\ ba = a^5 b\ (= a^{-1} b) \rangle$$

or as the group of symmetries of a regular hexagon. Both the generator–relation and the geometric view are useful.

The alternating group A_4 consists of the twelve even permutations in S_4:

e	(of order 1)
$(123), (132), (124), (142), (134), (143), (234), (243)$	(of order 3)
$(12)(34), (13)(24), (14)(23)$	(of order 2)

Our approach to classifying groups of order 12 will be to use the Sylow theorems. In order to give you an idea what to expect, we ask you to investigate the Sylow subgroups of the four groups of order 12 described above.

Exercise 3.1

For each of the groups given below, find their Sylow 2-subgroups and their Sylow 3-subgroups:

(a) $\mathbb{Z}_{12}$;

(b) $\mathbb{Z}_2 \times \mathbb{Z}_6$;

(c) D_6;

(d) A_4.

We now turn to the problem of classifying groups of order 12. That is, we assume that we have a group G, of order 12, and we show that it must be isomorphic to one of a list of 'known' groups. We already know that this list must contain at least four groups, those discussed in Exercise 3.1. In the course of our investigations we shall discover one further group that has to be added to the list.

Expanding on the remarks at the beginning of Solution 3.1, we know that G has either one or three Sylow 2-subgroups. Since they have order 4, each is isomorphic to $\mathbb{Z}_4$ or to the Klein group $V \cong \mathbb{Z}_2 \times \mathbb{Z}_2$. Similarly, G has either one or four Sylow 3-subgroups, each of which is isomorphic to $\mathbb{Z}_3$.

Since Sylow 2-subgroups are all conjugate, hence isomorphic, for a given G, they are all $\mathbb{Z}_4$ or all $\mathbb{Z}_2 \times \mathbb{Z}_2$.

There appear to be eight possibilities, which we summarize in the following table.

Case	Number of 2-subgroups	Number of 3-subgroups	Type of 2-subgroups	Type of 3-subgroups
1	1	1	$\mathbb{Z}_4$	$\mathbb{Z}_3$
2	1	1	$\mathbb{Z}_2 \times \mathbb{Z}_2$	$\mathbb{Z}_3$
3	1	4	$\mathbb{Z}_4$	$\mathbb{Z}_3$
4	1	4	$\mathbb{Z}_2 \times \mathbb{Z}_2$	$\mathbb{Z}_3$
5	3	1	$\mathbb{Z}_4$	$\mathbb{Z}_3$
6	3	1	$\mathbb{Z}_2 \times \mathbb{Z}_2$	$\mathbb{Z}_3$
7	3	4	$\mathbb{Z}_4$	$\mathbb{Z}_3$
8	3	4	$\mathbb{Z}_2 \times \mathbb{Z}_2$	$\mathbb{Z}_3$

To achieve our classification, we need a result about groups of permutations. It is a simple result, but surprisingly useful.

Lemma 3.1

Let G be a group of permutations. Then either *exactly half* of the permutations in G are even or they are *all* even.

Note that G need not be the whole of S_n.

Proof

Define a function ϕ from G to $\mathbb{Z}_2$ as follows:

$\phi : G \to \mathbb{Z}_2$
$\phi(g) = 0$ if g is even
$\phi(g) = 1$ if g is odd

Because the composite of even permutations is even, of odd permutations is even, and of an even permutation and an odd permutation (in either order) is odd, ϕ is a homomorphism.

Now, $\operatorname{Ker}(\phi)$ is the set of even permutations in G. Also, by the First Isomorphism Theorem,

$G / \operatorname{Ker}(\phi) \cong \operatorname{Im}(\phi)$.

The only possibilities for $\operatorname{Im}(\phi)$ are $\{0\}$ and $\mathbb{Z}_2$. So, either

$|G / \operatorname{Ker}(\phi)| = 1$,

in which case $|G| = |\operatorname{Ker}(\phi)|$ and all the permutations in G are even, or

$|G / \operatorname{Ker}(\phi)| = 2$,

in which case $|G| = 2 |\operatorname{Ker}(\phi)|$ and exactly half the permutations in G are even. ∎

You should now listen to the audio programme for this unit, referring to the tape frames below when asked to do so during the programme.

1 The problem

What are the groups of order 12?

2 Sylow possibilities

Case	Number of 2-subgroups	Number of 3-subgroups
1	1 ($\mathbb{Z}_4$)	1
2	1 ($\mathbb{Z}_2 \times \mathbb{Z}_2$)	1
3	1 ($\mathbb{Z}_4$)	4
4	1 ($\mathbb{Z}_2 \times \mathbb{Z}_2$)	4
5	3 ($\mathbb{Z}_4$)	1
6	3 ($\mathbb{Z}_2 \times \mathbb{Z}_2$)	1
7	3 ($\mathbb{Z}_4$)	4
8	3 ($\mathbb{Z}_2 \times \mathbb{Z}_2$)	4

($\mathbb{Z}_4$ or $\mathbb{Z}_2 \times \mathbb{Z}_2$)   (always $\mathbb{Z}_3$)

3 What we need to do

(a) Decide whether each case defines some group

(b) Find out whether any case defines more than one group

4 Known groups of order 12

| Abelian | $\mathbb{Z}_{12}$ | $\mathbb{Z}_2 \times \mathbb{Z}_6$ |
| Non-Abelian | D_6 | A_4 |

D_6 — dihedral group

A_4 — even permutations of S_4

5 Sylow structures of known groups

Group	2-subgroups	3-subgroups	Case
$\mathbb{Z}_{12}$	1 ($\mathbb{Z}_4$)	1 ($\mathbb{Z}_3$)	1
$\mathbb{Z}_2 \times \mathbb{Z}_6$	1 ($\mathbb{Z}_2 \times \mathbb{Z}_2$)	1 ($\mathbb{Z}_3$)	2
D_6	3 ($\mathbb{Z}_2 \times \mathbb{Z}_2$)	1 ($\mathbb{Z}_3$)	6
A_4	1 ($\mathbb{Z}_2 \times \mathbb{Z}_2$)	4 ($\mathbb{Z}_3$)	4

6. Cases 1 and 2

Know
(a) Unique (normal) Sylow 2-subgroup H
(b) Unique (normal) Sylow 3-subgroup K
(c) Coprime orders: $|H| = 4$, $|K| = 3$
(d) Product is order of group: $|H| \times |K| = 12$

Conclusion Group is internal direct product:
$$G \cong H \times K$$

Case 1: $H = \mathbb{Z}_4$, $K = \mathbb{Z}_3$; $G \cong \mathbb{Z}_4 \times \mathbb{Z}_3 \cong \mathbb{Z}_{12}$

Case 2: $H = \mathbb{Z}_2 \times \mathbb{Z}_2$, $K = \mathbb{Z}_3$; $G \cong \mathbb{Z}_2 \times \mathbb{Z}_2 \times \mathbb{Z}_3 \cong \mathbb{Z}_2 \times \mathbb{Z}_6$

7. Cases 3 and 4

Know
(a) One Sylow 2-subgroup  ($\mathbb{Z}_4$ or $\mathbb{Z}_2 \times \mathbb{Z}_2$)
 (must be normal)

(b) Four Sylow 3-subgroups ($\mathbb{Z}_3$)

Sylow 3-subgroups H_1, H_2, H_3, H_4  (all conjugate)

8. A group action

Group: G ($|G| = 12$)
Set: $X = \{H_1, H_2, H_3, H_4\}$
Action: $g \wedge H_i = g H_i g^{-1}$

Defines a homomorphism
$$\emptyset : G \to S_4$$

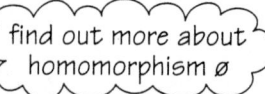 (find out more about homomorphism $\emptyset$)

9. Kernel of $\emptyset$

$g \in \text{Ker}(\emptyset)$ means the action of g produces the identity mapping on the set X

$g \in \text{Ker}(\emptyset)$ if and only if $g \in \text{Stab}(H_i)$, $i = 1, 2, 3, 4$

$$\text{Ker}(\emptyset) = \bigcap_{i=1}^{4} \text{Stab}(H_i)$$

10 Exercise 3.2

(a) Explain why $|\text{Orb}(H_i)| = 4$

(b) Use the Orbit–stabilizer Theorem to find $|\text{Stab}(H_i)|$

(c) Deduce that $\text{Stab}(H_i) = H_i$ *Hint* Show $H_i \subseteq \text{Stab}(H_i)$

(d) Show that $\text{Ker}(\phi) = \{e\}$

10A Solution 3.2

(a) Since all the H_is are conjugate, they belong to the same orbit

Hence $|\text{Orb}(H_i)| = 4$

(b) By the Orbit–stabilizer Theorem:
$$|G| = |\text{Orb}(H_i)| \times |\text{Stab}(H_i)|$$
$$12 = 4 \times |\text{Stab}(H_i)|$$

Hence $|\text{Stab}(H_i)| = 3$

(c) $h \in H_i$ implies $h \wedge H_i = h H_i h^{-1} = H_i$

So $H_i \subseteq \text{Stab}(H_i)$

But both have 3 elements, so
$$\text{Stab}(H_i) = H_i$$

(d) $\text{Ker}(\phi) = \bigcap_{i=1}^{4} \text{Stab}(H_i) = \bigcap_{i=1}^{4} H_i$

The H_is are distinct and of prime order, so their intersection is trivial

Hence $\text{Ker}(\phi) = \{e\}$

11 Consequences

By the First Isomorphism Theorem:
$$G/\text{Ker}(\phi) \cong \text{Im}(\phi)$$

Now $\text{Ker}(\phi) = \{e\}$, so
$$G \cong G/\text{Ker}(\phi)$$

Hence $G \cong \text{Im}(\phi) \subseteq S_4$

G is isomorphic to a 12-element subgroup of S_4

12

Exercise 3.3

(a) How many elements of order 3 does G have?

(b) What are the elements of order 3 in S_4? Are these elements odd or even?

(c) Deduce that Im(ø) has more than half of its elements even

(d) Conclude that Im(ø) = A_4

12A

Solution 3.3

(a) Each element of order 3 generates, and hence is in, a Sylow 3-subgroup

There are four such subgroups, each containing two elements of order 3

Hence G has eight elements of order 3

(b) The only elements of order 3 in S_4 are 3-cycles

The 3-cycles are even permutations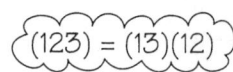

(c) The elements of order 3 in Im(ø) are 3-cycles, and so are even permutations

So Im(ø), of order 12, contains eight even permutations

(d) Since more than half the elements of Im(ø) are even, by Lemma 3.1 all its elements are even

Therefore Im(ø) consists of the twelve even elements of S_4, i.e. Im(ø) = A_4

13

Cases 3 and 4: conclusion

Case 3: cannot happen

Case 4: the group is A_4

14

Cases 5 and 6

Know   (a) Three Sylow 2-subgroups   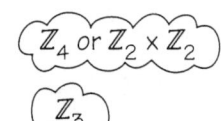

         (b) One Sylow 3-subgroup Z_3

Normal Sylow 3-subgroup $H = \{e, a, a^2\}$

15

Case 5

Let $K = \{e, b, b^2, b^3\}$ be *one* of the three cyclic Sylow 2-subgroups

Exercise 3.4

(a) Show that the right cosets
$H = He$, Hb, Hb^2 and Hb^3 are distinct

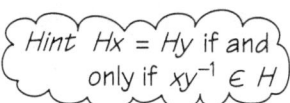
Hint $Hx = Hy$ if and only if $xy^{-1} \in H$

(b) What does this say about the twelve elements of G?

15A

Solution 3.4

(a) $Hb^i = Hb^j$ if and only if $b^i(b^j)^{-1} = b^{i-j} \in H$
Now $b^{i-j} \in H$ implies order of b^{i-j} is either 1 or 3
But $b^{i-j} \in K$ and so has order 1, 2 or 4
Hence b^{i-j} has order 1 and so $b^{i-j} = e$
So $b^i = b^j$ and He, Hb, Hb^2, Hb^3 are distinct

(b) As $H = \{e, a, a^2\}$, every element of G can be written as
$a^i b^j$ for $0 \leq i \leq 2$ and $0 \leq j \leq 3$

16

Products in standard form

Standard form: $a^i b^j$ $\quad 0 \leq i \leq 2, 0 \leq j \leq 3$
Product: $(a^i b^j)(a^k b^l) = a^? b^?$

Enough to be able to write
$$ba = a^? b^?$$

Know $H = \{e, a, a^2\}$ is normal in G, so that
$$bab^{-1} \in H$$

Exercise 3.5

Explain why $bab^{-1} = a^2$, i.e. $ba = a^2 b$

16A

Solution 3.5

Since $bab^{-1} \in H$, it is one of e, a, a^2

We eliminate the first two cases

If $bab^{-1} = e$ then $a = e$, contradicting the fact that a has order 3

If $bab^{-1} = a$ then $ba = ab$
Since a and b generate the group, this would imply the group is Abelian.
However, Abelian groups have unique Sylow subgroups,
contradicting our assumption about the Sylow 2-subgroups of G

The only remaining possibility is that $bab^{-1} = a^2$

(see Lemma 2.1)

17

Case 5: conclusion

Presentation:
$$G = \langle a, b : a^3 = e, b^4 = e, ba = a^2b \,(= a^{-1}b)\rangle$$

> new group

The elements of G are
$$a^i b^j, \quad 0 \leq i \leq 2, \, 0 \leq j \leq 3,$$
which are all distinct

18

Case 6

Let $K = \{e, b, c, bc\}$ be *one* of the three Klein Sylow 2-subgroups

Exercise 3.6

Show that the right cosets $H = He$, Hb, Hc and Hbc are distinct

> H is the Sylow 3-subgroup

18A

Solution 3.6

Let Hx and Hy be any of the given right cosets

Suppose $Hx = Hy$, so that $xy^{-1} \in H$

As $xy^{-1} \in H$, the order of xy^{-1} is 1 or 3

But $xy^{-1} \in K$, so the order of xy^{-1} is 1 or 2

So xy^{-1} has order 1, and therefore $x = y$

Hence the right cosets are distinct

19

Case 6: the generators

> $H = \{e, a, a^2\}$
> $K = \{e, b, c, bc\}$

Three generators: a, b, c

Standard form: $a^i b^j c^k \quad 0 \leq i \leq 2, \, j = 0, 1, \, k = 0, 1$

Relation: $bc = cb$

> K is Klein group
> b and c could be any two non-identity elements of K

If $ba = ab$ and $ca = ac$ then $(bc)a = a(bc)$, and the group is Abelian

A contradiction since G has 3 Sylow 2-subgroups

> Abelian groups have unique Sylow subgroups

Assume: $ba \neq ab$, i.e. $bab^{-1} \neq a$
$ca \neq ac$, i.e. $cac^{-1} \neq a$

Normality of H forces: $bab^{-1} = a^2$, i.e. $ba = a^2 b \,(= a^{-1}b)$
$cac^{-1} = a^2$, i.e. $ca = a^2 c \,(= a^{-1}c)$

20

Case 6: conclusion

Presentation:
$$G = \langle a, b, c : a^3 = e, b^2 = e, c^2 = e, ba = a^2b, ca = a^2c, bc = cb \rangle$$

The elements of G are
$$a^i b^j c^k, \quad 0 \leq i \leq 2, \quad j = 0, 1, \quad k = 0, 1,$$

which are all distinct

This group must be D_6

post-tape work shows how to derive usual presentation for D_6

21

Cases 7 and 8

Know (a) Three Sylow 2-subgroups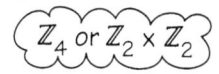

(b) Four Sylow 3-subgroups

Exercise 3.7

(a) How many elements of order 3 does G contain?

(b) Show that two *distinct* Sylow 2-subgroups contain at least five non-identity elements

(c) Deduce that no group corresponds to cases 7 or 8

21A

Solution 3.7

(a) Each of the four Sylow 3-subgroups has two elements of order 3. Hence G must have eight elements of order 3

(b) The intersection of two distinct Sylow 2-subgroups contains either one or two elements

If the intersection contains one element (the identity), then in total there are six non-identity elements

If the intersection contains two elements (the identity and one other), then each subgroup contains two further non-identity elements, and in total there are five non-identity elements

(c) The arguments above require the group (of order 12) to have at least thirteen non-identity elements

This contradiction shows there are no groups corresponding to these cases

Alternative proof

By a previous argument, if there are four Sylow 3-subgroups then the group is A_4

However, A_4 does not have three Sylow 2-subgroups — a contradiction

22 Groups of order 12

Case
1. $\mathbb{Z}_{12}$
2. $\mathbb{Z}_2 \times \mathbb{Z}_6$
4. A_4
5. $\langle a, b : a^3 = e, b^4 = e, ba = a^2b \,(= a^{-1}b) \rangle$
6. D_6

Case 5 dicyclic group of order 12

Let us now summarize the results from the tape. We began with the following table of possibilities for the Sylow subgroup structure of a group of order 12.

Case	Number of 2-subgroups	Number of 3-subgroups	Type of 2-subgroups	Type of 3-subgroups
1	1	1	$\mathbb{Z}_4$	$\mathbb{Z}_3$
2	1	1	$\mathbb{Z}_2 \times \mathbb{Z}_2$	$\mathbb{Z}_3$
3	1	4	$\mathbb{Z}_4$	$\mathbb{Z}_3$
4	1	4	$\mathbb{Z}_2 \times \mathbb{Z}_2$	$\mathbb{Z}_3$
5	3	1	$\mathbb{Z}_4$	$\mathbb{Z}_3$
6	3	1	$\mathbb{Z}_2 \times \mathbb{Z}_2$	$\mathbb{Z}_3$
7	3	4	$\mathbb{Z}_4$	$\mathbb{Z}_3$
8	3	4	$\mathbb{Z}_2 \times \mathbb{Z}_2$	$\mathbb{Z}_3$

Our conclusions were as follows:

1. This case gives the Abelian group $\mathbb{Z}_{12}$.
2. This case gives the Abelian group $\mathbb{Z}_2 \times \mathbb{Z}_6$.
3. This case cannot occur.
4. This case gives the alternating group A_4.
5. This case gives the dicyclic group of order 12:
$$\langle a, b : a^3 = e, \, b^4 = e, \, ba = a^2b \,(= a^{-1}b)\rangle.$$
6. This case gives the dihedral group D_6.
7. This case cannot occur.
8. This case cannot occur.

There are two outstanding matters. First, we need to show that the presentation given in Frame 20 is indeed that of D_6. Second, in four of the five cases that do occur, we know that the groups exist since we have a concrete example in each case; but this is not so for the dicyclic group. All that we have done is to show that if such a group exists, then it must have the presentation displayed above; so we need a concrete example.

We ask you, in the following exercise, to show that the presentation given in Frame 20 does indeed correspond to D_6.

Exercise 3.8

Show that the presentation
$$G = \langle a, b, c : a^3 = e, \ b^2 = e, \ c^2 = e, \ ba = a^2 b, \ ca = a^2 c, \ bc = cb \rangle$$
represents the same group as
$$D_6 = \langle a, b : a^6 = e, \ b^2 = e, \ ba = a^5 b \ (= a^{-1}b) \rangle.$$

Hint Consider the element abc.

Now we turn to the task of finding a concrete example of the dicyclic group of order 12.

The given presentation certainly defines a group. However, we cannot be sure, from the presentation, that it defines a group of order 12. If it does exist, it is certainly a new group of order 12 since it is non-Abelian and has an element b of order 4. (The other non-Abelian groups above, namely A_4 and D_6, do not have elements of order 4.)

The work we did in *Unit IB4* guarantees the existence of a group corresponding to any finite presentation that we care to write down. It does not give any guarantees about the *order* of the group.

So, to complete our investigation of groups of order 12, we need a concrete example of a group of order 12 having the following presentation:
$$\langle a, b : a^3 = e, \ b^4 = e, \ ba = a^2 b \ (= a^{-1}b) \rangle.$$

Before going on to find such a concrete example, we remark that the presentation defines a group of order *at most* 12. This is because the relation
$$ba = a^2 b$$
allows us to write any word in the generators a and b with all the as at the front. The relations $a^3 = e$ and $b^4 = e$ allow all such words to be written in the standard form
$$a^i b^j, \quad i = 0, \ldots, 2, \ j = 0, \ldots, 3.$$

Furthermore, the relations guarantee that this set of elements is closed under multiplication and also guarantee that every element in the set has an inverse which is also in the set.

The only possible remaining problem is that, though there are 12 such expressions of the form $a^i b^j$, we cannot (yet) be sure that they are all distinct. In the following exercise we give a concrete example of a group satisfying the relations above and ask you to verify that it does contain 12 elements. The result of this exercise will verify that there is a dicyclic group of order 12.

Exercise 3.9

Let
$$a = (123) \quad \text{and} \quad b = (12)(4567)$$
be elements of S_7.

(a) Show that a has order 3, b has order 4 and that
$$ba = a^2 b.$$

(b) List the 12 elements
$$a^i b^j, \quad i = 0, \ldots, 2, \ j = 0, \ldots, 3$$
in cycle form and hence show that they are all distinct.

From the solution to Exercise 3.9, the set of elements
$$\{e, (12)(4567), (46)(57), (12)(4765), (123), (13)(4567), (123)(46)(57),$$
$$(13)(4765), (132), (23)(4567), (132)(46)(57), (23)(4765)\}$$
is a 12-element subgroup of S_7 satisfying the presentation for the dicyclic group of order 12. So the dicyclic group of order 12 exists.

It can also be seen that this group has three Sylow 2-subgroups, each isomorphic to $\mathbb{Z}_4$, namely

$$\{e, (12)(4567), (46)(57), (12)(4765)\},$$
$$\{e, (13)(4567), (46)(57), (13)(4765)\},$$
$$\{e, (23)(4567), (46)(57), (23)(4765)\},$$

and just one Sylow 3-subgroup

$$\{e, (123), (132)\}.$$

This accounts for 10 of the 12 elements. The other two are $(123)(46)(57)$ and $(132)(46)(57)$, which are of order 6.

In fact, the presentation of this group given above is not the standard one for the dicyclic group of order 12. The standard presentation is

$$\langle c, d : c^6 = e, \ c^3 = d^2, \ dc = c^5 d \ (= c^{-1} d) \rangle.$$

From this standard presentation, however, it is by no means clear that the group so presented has 12 elements.
Certainly the second relation implies that

$$d^4 = (c^3)^2 = c^6 = e.$$

So we can argue as before that all elements of the group may be written in the form

$$c^i d^j, \quad i = 0, \ldots, 5, \ j = 0, \ldots, 3.$$

We now replace every d^2 by c^3 and reduce the resulting power of c modulo 6. This shows that every element of the group is of the form

$$c^i d^j, \quad i = 0, \ldots, 5, \ j = 0, 1.$$

We are now in the same position as we were with the original presentation of the dicyclic group. We have a presentation of a group which is of order at most 12.
If you wish to try it, the following exercise gives a concrete example of a group with presentation

$$\langle c, d : c^6 = e, \ c^3 = d^2, \ dc = c^5 d \ (= c^{-1} d) \rangle,$$

which turns out to be the concrete example of the dicyclic group of order 12 given in Exercise 3.9.

Exercise 3.10

Let

$$c = (123)(46)(57) \quad \text{and} \quad d = (12)(4567)$$

be elements of S_7.

(a) Show that c has order 6, d has order 4 and that

$$c^3 = d^2 \quad \text{and} \quad dc = c^5 d \ (= c^{-1} d).$$

(b) List the 12 elements

$$c^i d^j, \quad i = 0, \ldots, 5, \ j = 0, 1$$

in cycle form and hence show that they are all distinct.

The reason for introducing the standard presentation of the dicyclic group of order 12 is that this is the one which generalizes to produce a whole class of dicyclic groups, one for each integer of the form $4m$, for any positive integer m. The **dicyclic group of order $4m$** has presentation

$$\langle c, d : c^{2m} = e, \ c^m = d^2, \ dc = c^{2m-1} d \ (= c^{-1} d) \rangle,$$

which is a generalization of the standard presentation given for the dicyclic group of order 12 (given by taking $m = 3$).

When $m = 1$, the dicyclic group of order 4 just turns out to be the cyclic group of order 4. When $m = 2$, the dicyclic group of order 8 is the quaternion group. For larger values of m, the presentations define a whole new set of groups which you may meet if you study group theory further.

4 PRACTICE EXERCISES

This section, consisting entirely of exercises, is intended to give you some further practice in the techniques from the Groups stream.

Exercise 4.1

The dihedral group D_4 is given by the presentation

$$D_4 = \langle r, s : r^4 = e,\ s^2 = e,\ sr = r^3 s \left(= r^{-1} s\right)\rangle,$$

and the elements are written in standard form as

$$r^m \text{ and } r^m s, \quad m = 0, \ldots, 3.$$

(a) Express the product

$$(rs)\left(r^2 s\right)$$

in standard form.

(b) Express $\left(r^3 s\right)^{-1}$ in standard form.

Exercise 4.2

Let G be a group and let H and K be normal subgroups of G.
Prove that $H \times K$ is a normal subgroup of the group $G \times G$.

Exercise 4.3

Let G be the group

$$\mathbb{Z}_6 \times \mathbb{Z}_8.$$

Find the subgroups of G having order 4, and list the elements of each.
For each such subgroup, state whether it is isomorphic to $\mathbb{Z}_4$ or to $\mathbb{Z}_2 \times \mathbb{Z}_2$.

Exercise 4.4

The finitely presented Abelian group A is defined by the presentation

$$A = \langle a, b : 12a + 6b = 0,\ 3a + 2b = 0 \rangle.$$

(a) Write down the integer matrix representing A and use the Reduction Algorithm to reduce it to diagonal form.

(b) Express A as a direct product of cyclic groups in canonical form.

Exercise 4.5

Classify all Abelian groups of order 180 by writing each as a direct product of cyclic groups in canonical form.

5 WHERE NOW?

After expending a lot of effort studying a topic, it is often slightly discouraging to find out that one has only just scratched the surface. However, that really is all that we have been able to do in this course.

None the less, in spite of the wide scope of group theory, there are a number of areas that can reasonably be thought of as logical 'next steps' from what we have been able to discuss in the Groups stream.

This final section has a rather different purpose from the other sections of this unit. It gives an indication of what ideas build directly on the work in the Groups stream. We have concentrated on some themes that you might wish to look for in further reading, rather than recommending particular books. This is because the life of the course is likely to be long enough for some current texts to be out of print before the course is.

We do not expect you to spend much time on the details, and there are no exercises. Also, we restrict ourselves to *finite* groups unless we explicitly say otherwise.

5.1 Composition series

If A is an Abelian group of order $n > 1$ then either n is prime or it has a positive divisor not equal to itself or 1.
If n is prime, then A has no non-trivial proper subgroups, and so is cyclic of prime order.
If n is *not* prime, then there is a prime, p say, which divides n. Suppose that $n = pk$. Then A has a subgroup A_1 of order k which is automatically normal in A; moreover,

$$|A/A_1| = n/k = p.$$

Hence, the quotient group A/A_1 is cyclic of order p.

Now we repeat the argument with the subgroup A_1. We find that either $A_1 = \{e\}$ or A_1 is cyclic of prime order or we obtain a subgroup A_2 such that the quotient group

$$A_1/A_2$$

is cyclic of prime order. We continue this process. The orders of the subgroups we obtain are decreasing, so eventually we must arrive at the trivial subgroup $\{e\}$.

The second prime may well be different from the first.

Thus we have, for an Abelian group A, a *series* of subgroups

$$\{e\} = A_m \subset A_{m-1} \subset \cdots A_1 \subset A_0 = A$$

with the property that each subgroup is normal in the one above and all the quotient groups

$$A_i/A_{i+1}, \quad i = 0, \ldots, m - 1$$

are cyclic of prime order.

A reasonable question is whether this process can be extended to non-Abelian groups. The answer is a definite 'no'. The alternating group A_5 has 60 elements, and 60 is not prime. However, A_5 has no non-trivial, proper, normal subgroups at all, i.e. it is a simple group, so the process cannot even get started. The only series of normal subgroups for A_5 is

$$\{e\} \subset A_5,$$

We indicate below how to justify this assertion.

and the quotient group $A_5/\{e\} \cong A_5$ is not cyclic of prime order.

However, something can be rescued. Suppose that G is a finite group. If possible, find a non-trivial, proper, normal subgroup N of G, so that we have

$$\{e\} \subset N \subset G.$$

We may be able to insert further normal subgroups in two places: 'below' N or 'above' N. We consider how this might be done for each case separately.

If N contains a non-trivial, proper, normal subgroup N', we insert it:

$$\{e\} \subset N' \subset N \subset G.$$

To insert a subgroup 'above' N we look at the quotient G/N. If this contains a non-trivial, proper, normal subgroup, then we use an idea discussed in *Unit GR4*, Section 5. Any such normal subgroup must be of the form

$$N''/N,$$

Correspondence Theorem (Theorem 5.2).

where N'' is a normal subgroup of G between N and G:

$$N \subset N'' \subset G.$$

We continue inserting subgroups in this way. That is, if we have arrived at a chain of proper subgroups, each normal in the one above, we can try to insert another such subgroup between any two members of the chain.

We end up with a series of subgroups

$$\{e\} = N_m \subset N_{m-1} \subset \cdots \subset N_1 \subset N_0 = G$$

with the property that each subgroup is normal in the next and each quotient

$$N_i/N_{i+1}, \quad i = 0, \ldots, m-1,$$

has no non-trivial, proper, normal subgroups. Here, however, we cannot claim that the quotients are cyclic of prime order.

Each quotient is a simple group.

Such a series of proper normal subgroups is called a **composition series** for G.

Applying the method of construction indicated above to the permutation group S_3 gives a composition series

$$\{e\} \subset \{e, (123), (132)\} \subset S_3.$$

In this particular example the quotient groups have orders

$$6/3 = 2 \quad \text{and} \quad 3/1 = 3,$$

This example shows that a non-Abelian group can have a composition series with quotients of prime order.

and so they are cyclic of prime order.

One difficulty with composition series is that they are not necessarily unique, even for Abelian groups. For example, there is only one Abelian group A of order 15 and it is cyclic. It has subgroups H and K of orders 3 and 5 respectively. These give two different composition series:

$$\{0\} \subset H \subset A;$$
$$\{0\} \subset K \subset A.$$

The quotient groups in the first case are

$$H/\{0\} \cong H \cong \mathbb{Z}_3 \quad \text{and} \quad A/H \cong \mathbb{Z}_5.$$

In the second case they are

$$K/\{0\} \cong K \cong \mathbb{Z}_5 \quad \text{and} \quad A/K \cong \mathbb{Z}_3.$$

Although the series are different, the collection of quotient groups that we obtain is the same. It is this fact that generalizes. In most standard texts you will find a proof of this under the title of the *Jordan–Hölder Theorem*. It is about as close to an analogue of the Unique Prime Factorization Theorem as you can have for finite groups.

Unit GR1

We close this subsection by commenting on the assertion made earlier about the alternating group A_5, namely that it is simple. A straightforward, if lengthy, calculation would show that the class equation for A_5 is

$$60 = 1 + 12 + 12 + 15 + 20.$$

In *Unit GR4* we showed that a group with this class equation cannot have a non-trivial, proper, normal subgroup.

Exercise 4.10

5.2 Soluble and simple groups

In the discussion above, we observed that every finite Abelian group has a composition series with quotients which are cyclic of prime order. We also showed that S_3 had the same property. By considering A_5, we also showed that not all non-Abelian groups have this property. Thus, in some sense, the property

has a composition series with quotients of prime order

is a generalization of the Abelian property. It is a property of all Abelian groups and some non-Abelian ones.

A finite group which possesses a composition series with quotients of prime order is called a **soluble group**.

Thus, all finite Abelian groups are soluble, as is S_3, whereas A_5 is not soluble.

Historically, the solubility of groups was one of the first properties studied in detail. That was because the concept is intimately linked to the existence of formulae for solving polynomial equations. The fact that there exist formulae for solving any quadratic, cubic or quartic equation depends ultimately on the fact that S_2, S_3 and S_4 are all soluble. That no such formula exists for quintic equations is because S_5 is not soluble. This link, which is a reason for using the term 'soluble', was developed by Galois and you will find accounts in texts on Galois theory.

The unique composition series for S_5 is
$$\{e\} \subset A_5 \subset S_5.$$

The opposite extreme to solubility is, in a sense, exhibited by A_5: it has no non-trivial, proper, normal subgroups at all; that is, it is *simple*. A long-standing problem in group theory has been to classify all finite simple groups. We know that cyclic groups of prime order are simple because they have no non-trivial proper subgroups at all. They are also soluble, because $\mathbb{Z}_p$, p prime, has composition series

$$\{0\} \subset \mathbb{Z}_p$$

and the single quotient group is cyclic of prime order. Any other group cannot be both simple and soluble.

The classification of simple groups is a task of great complexity but it is firmly believed that it has been completed.

Why phrase the statement in this way? For theorems such as the Sylow theorems, the proofs are short enough and straightforward enough for large numbers of people to have checked and understood them. Better still, there are several different ways of proving them. Thus the degree of belief in the proofs approaches absolute certainty. For the classification of finite simple groups the proofs are of a quite different order of length. Relatively few people have the background necessary to do the checking or to search for alternative proofs. (As an example, the proofs required to show that all groups of odd order are soluble run to something like 1000 pages.)

5.3 Nilpotent groups

For Abelian groups we obtained an alternative to the canonical decomposition, the primary decomposition, in which the Abelian group is expressed as a direct product of groups of prime power order. Thus, if the Abelian group A has order n with prime decomposition

$$n = p_1^{\alpha_1} \ldots p_k^{\alpha_k},$$

then A can be written

$$A = P_1 \times \cdots \times P_k$$

where

$$|P_i| = p_i^{\alpha_i}, \quad i = 1, \ldots, k.$$

What this does, effectively, is to write A as the direct product of its Sylow subgroups.

There are non-Abelian groups which can be written as the direct product of their Sylow subgroups. As a slightly artificial example, consider the group G defined by

$$G = D_4 \times \mathbb{Z}_3.$$

Since G has order $8 \times 3 = 24$, and

$$24 = 2^3 \times 3,$$

its Sylow subgroups have orders 8 and 3. Thus D_4 is a Sylow 2-subgroup and $\mathbb{Z}_3$ is a Sylow 3-subgroup. However, by the properties of direct products, both are normal in G, hence, since Sylow p-subgroups are conjugate, they are unique. Thus, by the Internal Direct Product Theorem, G is the direct product of its Sylow subgroups. Because D_4 is non-Abelian, so is G.

A group which can be expressed as the direct product of its Sylow subgroups is called **nilpotent**.

> This is not the property usually taken as the definition. However, a standard theorem says that it is equivalent to the usual definition.

A nilpotent group must have the property that the Sylow subgroup for each distinct prime is normal and hence unique. Thus S_3 is not nilpotent because it has three Sylow 2-subgroups.

5.4 Representations

When we discussed the quaternion group earlier in the course, we gave a group of matrices as a concrete example. We started from a presentation, and a formal description of what we did is that we defined a homomorphism from Q to the set of all 2×2 non-singular real matrices. We defined it in such a way that Q was isomorphic to the image set.

> *Unit GR2*

A generalization of this idea is to ask what homomorphisms can exist from a given finite group to groups of matrices, and this question is the basis of what is called *representation theory*.

The modern approach tends to phrase the question slightly differently, in a way that links quite well with our extensive use of group actions. When we have used group actions we have ignored any algebraic structure that the set X being acted on may possess. Given a group G, we can ask what group actions we can define using G, if we insist that the set being acted on is a vector space. We also insist that the group action blends smoothly with the vector space structure. This means, for example, that if G acts on the real vector space V, we want

$$g \wedge (\alpha_1 \mathbf{v}_1 + \alpha_2 \mathbf{v}_2) = \alpha_1 (g \wedge \mathbf{v}_1) + \alpha_2 (g \wedge \mathbf{v}_2)$$

where $\mathbf{v}_1$ and $\mathbf{v}_2$ are vectors and α_1 and α_2 are real numbers.

You may well recognize this as saying that the action of g defines a linear transformation of V. As every linear transformation of V is associated (via a chosen basis) with a matrix, the action defines a homomorphism from G to a group of matrices.

Representation theory provides some attractive proofs to some quite difficult group-theoretic problems, just as group actions provided us with proofs of the Sylow theorems. It is true, however, that there is a school of thought which regards such proofs as 'cheating' in some degree and which prefers 'pure' group-theoretic methods.

5.5 Conclusion

We have tried to provide a number of pointers to areas of group theory beyond the scope of this course. Composition series and nilpotency are mainly of interest internally to group theory. Solubility is linked to Galois theory and is of general importance. Representation theory is of great importance in a number of applications. The one most closely linked to this course is in chemistry.

As the course has indicated, the classification of wallpaper patterns can be extended to three dimensions, and this classification is useful for determining possible crystal structures of solids. The idea of a symmetry group of a molecule, together with the representation theory of such symmetry groups, leads to important results about structure and stability of molecules.

SOLUTIONS TO THE EXERCISES

Solution 1.1

The number of Sylow 11-subgroups must be congruent to 1 modulo 11 and must divide 2. The only possibility is that there is 1 such subgroup.

The number of Sylow 2-subgroups must be congruent to 1 modulo 2 (i.e. odd) and must divide 11. There could be 1 or 11 such subgroups.

However, if there were only one subgroup of each order, G would have normal subgroups
$$H = \langle b \rangle \quad \text{and} \quad K = \langle a \rangle$$
of orders 2 and 11 respectively. Being of (different) prime orders, these subgroups would have trivial intersection.
Since K and bK would be the two distinct cosets of K in G, we have
$$G = HK.$$
Thus, G would be the internal direct product of the cyclic (and hence Abelian) subgroups H and K. This would imply that G is Abelian, contradicting the assumption that G is non-Abelian.

Hence G has 11 Sylow 2-subgroups.

Solution 1.2

The order of G is $405 = 3^4 \times 5$.

(a) The number of Sylow 3-subgroups is congruent to 1 modulo 3 and divides 5. The only possibility is that there is a unique Sylow 3-subgroup (of order 81).

The number of Sylow 5-subgroups is congruent to 1 modulo 5 and divides 81. The divisors of 81 are
$$1, 3, 9, 27, 81$$
of which only 1 and 81 are congruent to 1 modulo 5.

(b) Since, by the Sylow theorems, the H_is are conjugate, they belong to the same orbit, and so $|\text{Orb}(H_i)| = 81$. Therefore, we have
$$3^4 \times 5 = |G| = |\text{Orb}(H_i)| \times |\text{Stab}(H_i)| = 81 \times |\text{Stab}(H_i)|.$$
Hence $|\text{Stab}(H_i)| = 5$.

However, if $h \in H_i$, then
$$h H_i h^{-1} = H_i,$$
and so
$$H_i \subseteq \text{Stab}(H_i).$$
Since $|H_i| = 5 = |\text{Stab}(H_i)|$, we have
$$\text{Stab}(H_i) = H_i.$$

(c) Since
$$\text{Ker}(\phi) = \bigcap_{i=1}^{81} \text{Stab}(H_i) = \bigcap_{i=1}^{81} H_i,$$
and different Sylow 5-subgroups have trivial intersection, we have
$$\text{Ker}(\phi) = \{e\}.$$
Thus, ϕ is one–one and G is isomorphic to a subgroup of S_{81}.

Solution 2.1

(a) Since 2 is the highest power of 2 dividing the order of the group G, each Sylow 2-subgroup has order 2. Since 2 is a prime, the Sylow 2-subgroups are cyclic and isomorphic to $\mathbb{Z}_2$.

Similarly the Sylow p-subgroups are isomorphic to $\mathbb{Z}_p$.

(b) The number of Sylow p-subgroups is congruent to 1 modulo p and divides 2. Since $p > 2$ this implies that G has only one Sylow p-subgroup.

Solution 2.2

(a) For a group G of order qp where q and p are primes, the Sylow q-subgroups are isomorphic to $\mathbb{Z}_q$ and the Sylow p-subgroups are isomorphic to $\mathbb{Z}_p$.

(b) The number of Sylow p-subgroups is congruent to 1 modulo p and divides q. Since $q < p$ this implies that G has only one Sylow p-subgroup.

Solution 2.3

(a) The number of Sylow 2-subgroups is congruent to 1 modulo 2 and divides p. Since p is a prime, the only possibilities for the number of such subgroups is 1 or p.

(b) The only two groups of order 6 are $\mathbb{Z}_6$ and $S_3 \cong D_3$.

Since $\mathbb{Z}_6$ is cyclic it has a unique subgroup corresponding to each divisor of its order. Hence it has just one Sylow 2-subgroup.

Alternatively, the result follows since the group is Abelian, because this implies that every subgroup is normal. As a consequence, since, for any prime divisor p of the order of a group, the Sylow p-subgroups of the group are conjugate, all Sylow subgroups of $\mathbb{Z}_6$ are unique.

The group S_3 is

$$\{e, (12), (13), (23), (123), (132)\}$$

and so there are three Sylow 2-subgroups, namely

$$\{e, (12)\}, \quad \{e, (13)\} \quad \text{and} \quad \{e, (23)\}.$$

Solution 2.4

(a) Any group G of order $15 = 3 \times 5$ has, by Exercise 2.2, a unique (and hence normal) Sylow 5-subgroup, which is isomorphic to $\mathbb{Z}_5$.

The number of Sylow 3-subgroups is a divisor of 5 and so is either 1 or 5. It is also congruent to 1 modulo 3. As 5 is not congruent to 1 modulo 3, it follows that G has a unique (and hence normal) Sylow 3-subgroup isomorphic to $\mathbb{Z}_3$.

Thus, G is isomorphic to the internal direct product of its two normal Sylow subgroups, and so

$$G \cong \mathbb{Z}_3 \times \mathbb{Z}_5 \cong \mathbb{Z}_{15}.$$

Hence $\mathbb{Z}_{15}$ is the only group of order 15.

(b) The argument is exactly as above. A group G of order $77 = 7 \times 11$ has a unique (and hence normal) subgroup of order 11 (the larger prime factor). Further, the number of Sylow 7-subgroups is either one or eleven and, as eleven is not congruent to 1 modulo 7, it has one. So G has a unique (and hence normal) subgroup isomorphic to $\mathbb{Z}_7$. So

$$G \cong \mathbb{Z}_7 \times \mathbb{Z}_{11} \cong \mathbb{Z}_{77},$$

and $\mathbb{Z}_{77}$ is the only group of order 77.

Solution 2.5

If $i = 0$ then we have

$$bab^{-1} = e,$$

which implies that $a = e$, contradicting the fact that a has order p.

Solution 2.6

Since

$$bab^{-1} = a^i,$$

it follows that

$$b(bab^{-1})b^{-1} = ba^i b^{-1} = (bab^{-1})^i = (a^i)^i = a^{i^2}.$$

Now using the fact that b has order 2 gives

$$b(bab^{-1})b^{-1} = a = a^{i^2}.$$

Solution 2.7

From the previous exercise we have that

$$a^{i^2 - 1} = e.$$

Since a has order p, this implies that

$$p \mid (i^2 - 1), \quad \text{that is} \quad p \mid (i-1)(i+1).$$

Since p is prime this implies that either p divides $i - 1$ or p divides $i + 1$. The fact that $1 \leq i \leq p - 1$ reduces these two situations to

$$i = 1 \quad \text{or} \quad i = p - 1.$$

Solution 2.8

For $i = 1$ we have $bab^{-1} = a^1$, or equivalently

$$ba = ab.$$

Since K has order p and since bK and K are distinct, it follows that the elements a and b generate G, because the cosets K and bK account for all $2p$ elements of G. Hence the generators of G commute, and it follows that G is Abelian.

Since $|G| = 2p$, we have in this case that

$$G \cong \mathbb{Z}_2 \times \mathbb{Z}_p \cong \mathbb{Z}_{2p}.$$

For $i = p - 1$, we have $bab^{-1} = a^{p-1}$, or equivalently

$$ba = a^{p-1}b \, (= a^{-1}b).$$

Since a has order p and b has order 2, the resulting group has presentation

$$\langle a, b : a^p = e, \, b^2 = e, \, ba = a^{p-1}b \, (= a^{-1}b) \rangle,$$

which is a presentation for the dihedral group D_p.

Solution 3.1

Since $12 = 2^2 \times 3$, the Sylow 2-subgroups have order 4 and the Sylow 3-subgroups have order 3. Using the divisibility properties of the number of Sylow subgroups, there are either 1 or 3 Sylow 2-subgroups and either 1 or 4 Sylow 3-subgroups.

(a) Since $\mathbb{Z}_{12}$ is cyclic, it has a unique cyclic subgroup corresponding to each divisor of its order.

Hence the unique Sylow 2-subgroup of order 4 is

$$\{0, 3, 6, 9\} \cong \mathbb{Z}_4$$

and the unique Sylow 3-subgroup of order 3 is

$$\{0, 4, 8\} \cong \mathbb{Z}_3.$$

(b) Since $\mathbb{Z}_2 \times \mathbb{Z}_6$ is Abelian, all subgroups are normal, and hence there is a unique Sylow 2-subgroup and a unique Sylow 3-subgroup.

The Sylow 2-subgroup is

$$\{(0,0), (0,3), (1,0), (1,3)\} \cong \mathbb{Z}_2 \times \mathbb{Z}_2 \cong V$$

and the Sylow 3-subgroup is

$$\{(0,0), (0,2), (0,4)\} \cong \mathbb{Z}_3.$$

(c) The Sylow 2-subgroups, being of order 4, can only contain elements of orders 1, 2 or 4. However, D_6 has no elements of order 4, so the Sylow 2-subgroups must be isomorphic to the Klein group, V.

Thinking geometrically, the only elements of order 2 are the six reflections and the half-turn. In algebraic terms these are

$$b, ab, a^2b, a^3b, a^4b, a^5b, a^3.$$

Each Klein group has three elements of order 2, so D_6 must contain more than one Sylow 2-subgroup. Hence, by the remark at the start of this solution, there must be 3 such subgroups.

Any subgroup which contains two reflections must contain their product, which is a rotation through twice the angle between the reflection lines. Therefore, each of the three Sylow 2-subgroups consists of the identity, two reflections in lines at right angles and the half-turn. In terms of the generators, the subgroups are:

$$\{e, b, a^3b, a^3\};$$
$$\{e, ab, a^4b, a^3\};$$
$$\{e, a^2b, a^5b, a^3\}.$$

A Sylow 3-subgroup is cyclic and so has the identity and two elements of order 3. As the only elements of order 3 in D_6 are the rotations through $2\pi/3$ and $4\pi/3$, there is only one Sylow 3-subgroup. In terms of the generators it is

$$\{e, a^2, a^4\},$$

since a is a rotation through $2\pi/6$.

(d) The elements of A_4 have orders 1, 2 and 3. Arguing as for D_6, any Sylow 2-subgroup must be isomorphic to the Klein group and consists of the identity and three elements of order 2. Hence the unique Sylow 2-subgroup of A_4 is

$$\{e, (12)(34), (13)(24), (14)(23)\}.$$

A Sylow 3-subgroup is cyclic and consists of the identity and two elements of order 3, each being the inverse of the other. Hence there are four Sylow 3-subgroups:

$$\{e, (123), (132)\};$$
$$\{e, (124), (142)\};$$
$$\{e, (134), (143)\};$$
$$\{e, (234), (243)\}.$$

Solution 3.8

We know that D_6 has order 12. G also represents a group of order 12 since its twelve distinct elements are:

$$a^i b^j c^k, \quad i = 0, 1, 2, \ j = 0, 1, \ k = 0, 1.$$

Consider the element abc in G. We have:

$$(abc)^2 = a^2$$
$$(abc)^3 = bc$$
$$(abc)^4 = a$$
$$(abc)^5 = a^2 bc$$
$$(abc)^6 = e$$

We use the relations $a^3 = e$, $b^2 = e$, $c^2 = e$, $ba = a^2 b$, $ca = a^2 c$ and $bc = cb$ to derive these.

Also we have

$$b(abc) = a^2 c \quad \text{and} \quad (abc)^5 b = a^2 c,$$

so that

$$b(abc) = (abc)^5 b.$$

Thus G has 12 elements, an element abc of order 6, an element b of order 2 and a relation linking these two of the appropriate form. Hence it must be the dihedral group D_6.

In fact there is a one–one correspondence between the elements of G and those of D_6 as follows:

G		D_6		G		D_6
e	$\leftrightarrow$	e		bc	$\leftrightarrow$	a^3
a	$\leftrightarrow$	a^4		a^2	$\leftrightarrow$	a^2
b	$\leftrightarrow$	b		$a^2 b$	$\leftrightarrow$	$a^2 b$
c	$\leftrightarrow$	$a^3 b$		$a^2 c$	$\leftrightarrow$	$a^5 b$
ab	$\leftrightarrow$	$a^4 b$		abc	$\leftrightarrow$	a
ac	$\leftrightarrow$	ab		$a^2 bc$	$\leftrightarrow$	a^5

This correspondence can easily be used to check that the elements of D_6 satisfy the corresponding relations in G. (We've already seen that the elements of G satisfy the corresponding relations in D_6.)

Solution 3.9

(a) $\quad a = (123), \quad a^2 = (132) \quad \text{and} \quad a^3 = e.$

Hence a has order 3.

$$b = (12)(4567), \quad b^2 = (46)(57), \quad b^3 = (12)(4765) \quad \text{and} \quad b^4 = e.$$

Hence b has order 4.

$$ba = (12)(4567)(123) = (23)(4567);$$
$$a^2 b = (123)^2 (12)(4567) = (132)(12)(4567) = (23)(4567).$$

Hence

$$ba = a^2 b.$$

(b) Listing the 12 elements in disjoint cycle form gives the following:

$a^0 b^0 = e$
$a^0 b^1 = (12)(4567)$
$a^0 b^2 = (46)(57)$
$a^0 b^3 = (12)(4765)$
$a^1 b^0 = (123)$
$a^1 b^1 = (123)(12)(4567) = (13)(4567)$
$a^1 b^2 = (123)(46)(57)$
$a^1 b^3 = (123)(12)(4765) = (13)(4765)$
$a^2 b^0 = (132)$
$a^2 b^1 = (132)(12)(4567) = (23)(4567)$
$a^2 b^2 = (132)(46)(57)$
$a^2 b^3 = (132)(12)(4765) = (23)(4765)$

Hence the 12 elements
$$a^i b^j, \quad i = 0, \ldots, 2, \; j = 0, \ldots, 3$$
are distinct.

Solution 3.10

(a) The order of a product of disjoint cycles is the LCM of their lengths. So,
$$|c| = \mathrm{lcm}\{3, 2, 2\} = 6,$$
$$|d| = \mathrm{lcm}\{2, 4\} = 4.$$

Next,
$$c^3 = ((123)(46)(57))^3 = (123)^3 (46)^3 (57)^3 = (46)(57),$$
$$d^2 = ((12)(4567))^2 = (12)^2 (4567)^2 = (46)(57) = c^3.$$

We have
$$dc = (12)(4567)(123)(46)(57) = (23)(4765).$$

Now,
$$c^5 = c^{-1} = (123)^{-1}(46)^{-1}(57)^{-1} = (132)(46)(57),$$

so
$$c^5 d = c^{-1} d = (132)(46)(57)(12)(4567) = (23)(4765) = dc.$$

(b) Listing the 12 elements in disjoint cycle form gives the following:

$c^0 d^0 = e$
$c^0 d^1 = (12)(4567)$
$c^1 d^0 = (123)(46)(57)$
$c^1 d^1 = (13)(4765)$
$c^2 d^0 = (132)$
$c^2 d^1 = (132)(12)(4567) = (23)(4567)$
$c^3 d^0 = (46)(57)$
$c^3 d^1 = (46)(57)(12)(4567) = (12)(4765)$
$c^4 d^0 = (123)$
$c^4 d^1 = (123)(12)(4567) = (13)(4567)$
$c^5 d^0 = (132)(46)(57)$
$c^5 d^1 = (132)(46)(57)(12)(4567) = (23)(4765)$

Hence the 12 elements
$$c^i d^j, \quad i = 0, \ldots, 5, \; j = 0, 1$$
are distinct.

In fact, inspection of the previous solution shows that we have precisely the same concrete example of the dicyclic group of order 12.

As the exercises in Section 4 are designed to help your *revision* for the examination, the solutions that we have provided below contain more detail than would be strictly necessary for an examination solution.

Solution 4.1

(a) The basic technique involves using the relation $sr = r^3s$ as often as is necessary to bring all the rs to the front. Finally, we apply $r^4 = e$ and $s^2 = e$ if necessary.

$$(rs)(r^2s) = r(sr)rs$$
$$= r(r^3s)rs$$
$$= r^4(sr)s$$
$$= r^7s^2$$
$$= r^3.$$

Alternatively, using Theorem 1.1 of *Unit IB3*:
$$(rs)(r^2s) = r(sr^2)s$$
$$= r(r^6s)s$$
$$= r^7s^2$$
$$= r^3.$$

(b) The inverse of a product is the product of the inverses taken in the opposite order, i.e. $(ab)^{-1} = b^{-1}a^{-1}$.

$$(r^3s)^{-1} = s^{-1}(r^3)^{-1}$$
$$= sr \quad \text{(because } ss = e \text{ and } r^3r = e)$$
$$= r^3s.$$

Solution 4.2

All that we have to work with are the definitions of 'direct product' and 'normal subgroup'.

We must show both that $H \times K$ is a subgroup of $G \times G$ and that it satisfies the normality condition.

Subgroup

Closure

Let (h_1, k_1) and (h_2, k_2) be any two elements of the direct product $H \times K$. Then

$$(h_1, k_1)(h_2, k_2) = (h_1h_2, k_1k_2) \quad \text{(by the definition of } H \times K.\text{)}$$

However, since H is a subgroup, $h_1h_2 \in H$. Also, since K is a subgroup, $k_1k_2 \in K$. It follows, therefore, that

$$(h_1h_2, k_1k_2) \in H \times K.$$

Identity

The identity e of G belongs to both the subgroups H and K. It follows, therefore, that (e, e) is an element of $H \times K$. Furthermore, (e, e) is the identity of $G \times G$.

Inverses

Let (h, k) be any element of $H \times K$. Since H is a subgroup, $h^{-1} \in H$, and similarly $k^{-1} \in K$. Hence, the element (h^{-1}, k^{-1}), which is the inverse of (h, k), is in $H \times K$.

Normality property

Let (h,k) be any element in $H \times K$ and let (g_1, g_2) be any element of $G \times G$. Then

$$(g_1, g_2)(h,k)(g_1, g_2)^{-1} = (g_1, g_2)(h,k)(g_1^{-1}, g_2^{-1})$$
$$= (g_1 h g_1^{-1}, g_2 k g_2^{-1}).$$

Since H is a normal subgroup of G we have

$$g_1 h g_1^{-1} \in H.$$

Similarly

$$g_2 k g_2^{-1} \in K.$$

So

$$(g_1, g_2)(h,k)(g_1, g_2)^{-1} \in H \times K.$$

This completes the proof that $H \times K$ is a normal subgroup of $G \times G$.

Solution 4.3

A group of order 4 is either cyclic or a Klein group. We look for subgroups of G of each type in turn.

Cyclic

The order of the element (a,b) in $\mathbb{Z}_6 \times \mathbb{Z}_8$ is the lowest common multiple of the orders of a and b. We want (a,b) to have order 4.
Since $a \in \mathbb{Z}_6$, it has order 1, 2, 3 or 6. As we require

$$\mathrm{lcm}\{|a|, |b|\} = 4,$$

we can eliminate elements of orders 3 and 6.
Since $b \in \mathbb{Z}_8$, it has order 1, 2, 4 or 8. We can eliminate those of order 8 immediately. But as a has order 1 or 2 we can also eliminate those of orders 1 and 2.
We are left with a of order 1 or 2 and b of order 4.

Hence a is either 0 or 3 and b must be 2 or 6. Thus there are four elements of order four. The subgroups generated are:

$$\langle(0,2)\rangle = \{(0,2), (0,4), (0,6), (0,0)\} = \langle(0,6)\rangle;$$
$$\langle(3,2)\rangle = \{(3,2), (0,4), (3,6), (0,0)\} = \langle(3,6)\rangle.$$

Since these are cyclic of order four, both are isomorphic to $\mathbb{Z}_4$.

Klein

Every element of a Klein group has order 1 or 2. Thus, if (a,b) is an element of a Klein subgroup of $\mathbb{Z}_6 \times \mathbb{Z}_8$, both a and b must have order 1 or 2. Hence a is 0 or 3 and b is 0 or 4. There are, therefore, only four such elements and so only one Klein subgroup,

$$\{(0,0), (0,4), (3,0), (3,4)\},$$

which is isomorphic to $\mathbb{Z}_2 \times \mathbb{Z}_2$.

Solution 4.4

(a) The integer matrix representing A is

$$\begin{bmatrix} 12 & 6 \\ 3 & 2 \end{bmatrix}.$$

Using the Reduction Algorithm we obtain the following:

$$\begin{bmatrix} 12 & 6 \\ 3 & 2 \end{bmatrix} \mapsto \begin{bmatrix} 3 & 2 \\ 12 & 6 \end{bmatrix} \quad R_1 \leftrightarrow R_2$$

$$\mapsto \begin{bmatrix} 2 & 3 \\ 6 & 12 \end{bmatrix} \quad C_1 \leftrightarrow C_2$$

$$\mapsto \begin{bmatrix} 2 & 1 \\ 6 & 6 \end{bmatrix} \quad C_2 \to C_2 - C_1$$

$$\mapsto \begin{bmatrix} 1 & 2 \\ 6 & 6 \end{bmatrix} \quad C_1 \leftrightarrow C_2$$

$$\mapsto \begin{bmatrix} 1 & 2 \\ 0 & -6 \end{bmatrix} \quad R_2 \to R_2 - 6R_1$$

$$\mapsto \begin{bmatrix} 1 & 0 \\ 0 & -6 \end{bmatrix} \quad C_2 \to C_2 - 2C_1$$

$$\mapsto \begin{bmatrix} 1 & 0 \\ 0 & 6 \end{bmatrix} \quad R_2 \to -R_2$$

(b) From the diagonal form of the reduced matrix we have

$$A \cong \mathbb{Z}_1 \times \mathbb{Z}_6 \cong \mathbb{Z}_6.$$

This is already in canonical form.

There are many ways of reducing the original matrix to diagonal form. If you followed the Reduction Algorithm exactly then you will have got the form above. Some other combinations of elementary row and column operations lead to

$$\begin{bmatrix} 2 & 0 \\ 0 & 3 \end{bmatrix},$$

and hence to

$$A \cong \mathbb{Z}_2 \times \mathbb{Z}_3 \cong \mathbb{Z}_6.$$

Solution 4.5

First we observe that

$$180 = 2^2 \times 3^2 \times 5.$$

We now obtain the torsion coefficients by drawing up a table of possibilities.

prime power	factors of d_{k-1}	d_k	label
5		5	5a
3^2		3^2	3a
	3	3	3b
2^2		2^2	2a
	2	2	2b

Hence, there are $1 \times 2 \times 2 = 4$ Abelian groups of order 180. They are:

$\mathbb{Z}_{180}$ 5a, 3a, 2a
$\mathbb{Z}_2 \times \mathbb{Z}_{90}$ 5a, 3a, 2b
$\mathbb{Z}_3 \times \mathbb{Z}_{60}$ 5a, 3b, 2a
$\mathbb{Z}_6 \times \mathbb{Z}_{30}$ 5a, 3b, 2b

OBJECTIVES

After studying this unit, you should be able to:
(a) analyse possible Sylow structures for a group of a given order;
(b) classify groups of order $2p$, for p a prime greater than 2;
(c) classify certain groups of order qp, for distinct primes q and p;
(d) classify groups of a given (small) order in simple cases.

INDEX

composition series 28, 29
decomposition theorem for finitely
    generated Abelian groups 5
dicyclic group 26
dicyclic group of order $4m$ 26
dicyclic group of order 12 24
group action 7

groups of order $2p$ 11, 13
groups of order $3p$ 14
groups of order qp 13
groups of order 12 15, 24
homomorphism derived from group
    action 7
Jordan–Holder Theorem 29

nilpotent group 31
representation theory 31
simple group 30
soluble group 30